YOUR KNOWLEDGE HAS VALUE

- We will publish your bachelor's and master's thesis, essays and papers

- Your own eBook and book - sold worldwide in all relevant shops

- Earn money with each sale

Upload your text at www.GRIN.com and publish for free

The Earthworms Role in Sustainable Agricultural Practices and Vermicomposting

Mallikarjun Yadawade

Bibliographic information published by the German National Library:

The German National Library lists this publication in the National Bibliography; detailed bibliographic data are available on the Internet at http://dnb.dnb.de.

ISBN: 9783346983848
This book is also available as an ebook.

Print and binding: Books on Demand GmbH, Norderstedt, Germany
Printed on acid-free paper from responsible sources.

The present work has been carefully prepared. Nevertheless, authors and publishers do not incur liability for the correctness of information, notes, links and advice as well as any printing errors.

GRIN web shop: https://www.grin.com/document/1431257

Vermicompost: An Ecofriendly Fertilizer and Bio-Agent for Agricultural Crops; A Review

Mallikarjun S. Yadawade

Abstract

Vermicomposting is the process of turning organic waste into nutrient-rich compost by utilizing earthworms. The casting made by earthworms is nutritional organic manure that contains growth hormones, enzymes, humus, NPK, micronutrients, and beneficial microorganisms. Numerous researches in the literature demonstrate the advantages of incorporating vermicompost into soil. Applying vermicompost as organic manure to soil increased microbial activity, microbial biomass carbon, enzymatic activity, improved nutritional status, and increased cation exchange capacity. The castings made by earthworms also have anti-pest properties. In addition, vermicompost enhances soil aggregation, structure, and water-retention ability.

Keywords: vermicompost, soil health, earthworm, protection

Introduction

For more than 20 million years, earthworms have existed on our planet. Earthworms are soil-dwelling organisms that consume decomposing organic matter. Following digestion, the undigested material passes through the earthworm's alimentary canal and coats the castings in a thin layer of oil. Within a fortnight, this layer begins to degrade. So, even though the nutrients for the plant are instantly available, they are delivered gradually to make them last longer. Organic waste is converted to natural fertiliser in the earthworm's alimentary canal through a process. Organic wastes go through chemical transformations such as neutralisation and deodorization. This indicates that the castings have a pH of 7 (neutral) and have no odour (Tomati and Galli, 1995).

The term vermicomposting means the use of earthworms for composting organic residues. Earthworms can consume practically all kinds of organic matter and they can eat their own body weight per day, e.g. 1 kg of worms can consume 1 kg of residues every day. The excreta (castings) of the worms are rich in nitrate, available forms of P, K, Ca and Mg. The passage of soil through earthworms promotes the growth of bacteria and actinomycetes. Actinomycetes thrive in the presence of worms and their content in worm casts is more than six times that in the original soil (Edwards, 1998).

Types of Worms

There are more than 3000 species of earthworms in the soil (Cook and Linden, 1996), but hardly 8-10 species are found suitable for vermicompost preparation. The best types of earthworms for vermiculture and vermicomposting are epigaeic species (litter dwellers, live in organic horizon) such as *Eisenia fetida* and *Eudrilus eugeniae* (Dominguez and Edwards, 2004). A moist compost heap of 2.4 m by 1.2 m and 0.6 m high can support a population of more than 50 000 worms. The introduction of worms into a compost heap has been found to mix the materials, aerate the heap and hasten decomposition. Turning the heaps is not necessary where earthworms are present to do the mixing and aeration. The ideal environment for the worms is a shallow pit and the right sort of worm is necessary. *Lumbricus rubellus* (red worm) and *Eisenia foetida* are thermo-tolerant and so particularly useful. Field worms (*Allolobophora caliginosa*) and night crawlers (*Lumbricus terrestris*) attack organic matter from below but the latter do not thrive during active composting, being killed more easily than the others at high temperature (Bansal and Kapoor, 2000.).

European night crawlers (*Dendrabaena veneta* or *Eisenia hortensis*) are produced commercially and have been used successfully in most climates. These night crawlers grow to about 10–20 cm. The African night crawler (*Eudrilus eugeniae*), is a large, tropical worm species. It tolerates higher temperatures than *Eisenia foetida* does, provided there is ample humidity. However, it has a narrow temperature tolerance range, and it cannot survive at temperatures below 7 °C. Vermicomposting is in use in many countries.

The gizzard of earthworms breaks down organic waste physically. It is then exposed to several enzymes released into the lumen by the gut wall and related bacteria, including protease, cellulose, lipase, chitinase, and amylase. The aforementioned enzymes convert complicated biomolecules into simpler ones. Vermicompost's structural stability is provided by the mucus secretions from the gut wall. The remaining material is expelled by earthworms as casting, with just 5–10% of the material being absorbed for their growth (Kiehl, 2001).

Methods of vermicompost preparation

Site selection

In general, earthworms like to live in damp, shaded areas because these are ideal for their rapid growth. According to Rostami et al. (2009), worms are more restricted by high temperatures and dry environments than by low temperatures and saturated water. Because earthworms cannot survive in standing water, the vermicompost unit needs to have enough drainage. The ideal temperature range for the earthworms to multiply and grow more quickly is between 20 and 30 degrees Celsius. Therefore, high-quality compost can be quickly created in ambient temperature circumstances by employing the right type of earthworms. If possible, choose a damp, shaded area beneath the tree or beneath the ventilated shed. Vermicompost beds or units should have adequate drainage, and a water supply should be close to the unit. If not, earthworms will use the carbon in the biogas as food and slow down the breakdown of the material. It should be far from the biogas plant.

Anatomy of earthworm

The earthworm is spherical in shape, with a tall, pointed head and a somewhat flattened back end. The earthworm can twist and turn because of the rings that round its soft, fluid body—especially because it lacks a backbone. The earthworm can crawl because it lacks true legs and can move its bristles, or setae, back and forth on its body. The earthworm uses its skin as a breathing tube. Food enters the stomach (crop) through the mouth. Subsequently, the meal travels via the gizzard, where swallowed stones crush it. What remains is removed after going through the colon for digestion.

With both male and female sex organs, earthworms are hermaphrodites; nevertheless, in order to mate, they need another earthworm. A mature breeding earthworm secretes mucus (albumin) after mating, which is released via the wide band (clitellum) that surrounds it. Storage sacs contain sperm from another worm. Sperm and eggs are encased in the mucus as it travels over the worm. The cocoon forms a lemon-shaped shell that is about 1/8 inch long after both ends seal after escaping the worm. It takes around three weeks for one end of the cocoon to produce two or more baby worms. Worms are 1/2 to 1 inch long and pale to virtually translucent as babies. The sexual maturity of redworms takes 4–6 weeks (Allen, 2016).

Steps

One great way to lessen environmental risks and provide a healthy, natural soil supplement is by composting. Vermicompost is actually a fantastic substitute that permits composting to be done with the least amount of area needed. Vermicomposting has become a hot topic practically everywhere due to its simplicity and versatility in terms of application. Producing high-quality vermicompost from raw trash necessitates extensive process understanding. The many phases in the creation of vermicompost are described in more detail below (Subbulakshmi and Thiruneelakandan, 2011):

- ➤ Smoothen the surface once the location for the vermicompost preparation has been chosen. Now, use bricks to create a bed that is roughly $10 \times 3 \times 3$ feet (L x B x H). The bed's dimensions, however, might be altered depending on the amount of material required and available.

- ➤ Sprinkle some water on the bed to moisten its surface. Now, place a layer of dried leaves, paddy straw, etc. at the foot of the bed that is two to three inches thick.

- ➤ Over the layer of dried particles, scatter some water once more. Evenly cover the layer of leaves or straw with a layer of farm yard manure or cow dung, about one to one and a half feet thick, and spritz it with water to ensure it is sufficiently moist.

- ➤ The cow manure shouldn't be too recent. It needs to be at least ten to fifteen days old because freshly laid cow manure generates a lot of heat and can kill earthworms. In a similar vein, cow dung shouldn't be too old because it has decomposed and won't provide food for earthworms.

- ➤ Add the kitchen scraps now, slicing them into little bits, such as vegetable leaves, fruit rinds, grasses, and animal roughages. Once more, evenly distribute a layer of cow dung between one and three feet deep and use a sprinkler to provide enough water. Distribute one kilogramme of vermiculture, which contains around 800–1000 earthworms, onto the cow dung layer.

- ➤ Once more, evenly distribute a 2- to 3-inch layer of green leaves, etc. over the FYM layer and mist with water. Use jute or gunny sacks to cover the vermicompost bed at this point. Water the gunny bags with a sprinkler system on a regular basis to keep the vermicompost bed at the ideal moisture and temperature.

- The bed should be between 15 and 30 degrees Celsius with 35 to 40% moisture content. For this reason, water should be sprayed often to keep the ideal environment for earthworm development and activity. In order to preserve shade for the vermicompost unit and protect the earthworms from direct sunlight and precipitation, it is recommended to build a shed or roof over it if it is placed in an open area.
- In roughly eight to ten weeks, vermicompost can be made by following the aforementioned methods. When the vermicompost reaches maturity, it has a dark brown hue, is granulated, highly porous, and odourless.

Storage and harvesting

As soon as the vermicompost is ready, stop watering it (about a week ago) and pile the compost. The earthworms will now begin to descend and collect at the base of the pile. Take the top layer of material out of the pile and place it in the shade so that it can be sieved and packed later. A minimum of 40% moisture content is required, and direct sunlight should not hit heaped vermicompost as this could lead to moisture and nutrient loss.

After sieving the vermicompost, move any earthworms to the next bed. Many earthworms can be found in the lower part of the vermicompost heap, which could be utilised for vermiculture in order to prepare vermicompost once again. The vermicompost is now prepared for use in field, vegetable and fruit crops, flower pots, etc. If the vermicompost is kept at the ideal moisture level of 40%, it can be stored for a minimum of one year without losing any of its quality (Lores et al, 2006).

Characteristics of vermicompost

Vermicompost's acceptance in terms of sustainable crop production has been growing quickly as more people become aware of the importance of organic inputs in crop fields. Vermicompost, the primary substance that is regarded as earthworm excrement, possesses several attributes. These are as follows:

Physical characteristics

- A quality vermicompost is invariably non-toxic, thoroughly broken down, harmonious with the environment, and compatible with ecology.
- Earthworms can be utilised to convert any kind of green waste, including sewage sludge, industrial waste, agricultural waste, and human faeces.
- Turning soil properly is a sign of aerobic decomposition, which will result in a typical odour following preparation. Improper aeration may result in the formation of an unpleasant odour.
- The ultimate product of vermicomposting would be granular matter with fine particle structure.
- Vermicompost enhances the soil's porosity, drainage, and water-holding ability, acting as a "soil conditioner." (Edwards and Burrows, 1988).

Chemical characteristics

- Almost all necessary macro and micronutrients for plants are abundant in vermicompost.
- According to several experiments, vermicompost has an average nutrient content that is higher than that of other traditional compost made through other processes.
- Compared to other traditional compost, vermicompost contains worm mucus, which helps to keep nutrients from washing away. Of all the secondary nutrients, calcium content in vermicompost is higher than in other compost (Kalantari et al., 2010).
- Because heavy metal accumulates in worm tissue as a result of vermi-conversion, it is discovered that the amount of heavy metal found in feeding material is decreased in earthworm cast. The rate at which heavy metals are removed from feed depends on the vermicomposting technology used. Because of this characteristic, vermicompost is less contaminated than other composts. As a result, it gains greater environmental sustainability (Sahariah et al., 2015)

Biological characteristics

- Actinomycetes, fungus, and bacteria are among the microorganisms that live in the by-product of earth casting. Numerous enzymes and phytohormones are released by these microorganisms, which promotes better plant growth. Vermicompost thus promotes enzymatic and microbiological activity (Nada et al., 2011).

- The excreta of earthworms are thought to have a healthy range of populations of symbiotic associative bacteria and nitrogen fixer bacteria.
- Furthermore, a significant number of vesicular-arbuscular mycorrhiza (VAM) propagules are found in earthworm casts. These propagules can endure for up to 11 months on the cast, and they aid in boosting microbial activity to provide the plant with easily accessible forms of phosphate and nitrogen Reddell and Spain, 1991)

Nutritional value

The waste material or base substrate used to prepare the vermicompost usually determines how many nutrients it contains. According to Chowdeppa et al. (1999), the type of earthworms utilised in vermicomposting may also have an impact on the vermicompost's quality. According to Ushakumari et al. (1999), vermicompost made from banana wastes (leaves, pseudostem) and cattle manure in an 8:1 ratio typically included 1.5, 0.4, and 1.8% N, P2O5, and K2O, respectively. Similar to this, vermicompost made from various organic materials, such as leftover sugarcane, ipomea, parthenium, neem leaves, and banana peduncles, is very nutrient-rich, improving soil fertility and increasing rice productivity (Vasanthi and Kumaraswamy, 1999). Therefore, the foundation substrate utilised to produce vermicompost has a significant impact on its nutritional content.

Vermicompost and plant growth and development

Numerous plant species, particularly horticultural crops like sweet corn, tomatoes and strawberries (Arancon et al.,2004), cereal crops like rice (Tharmaraj et al., 2011), wheat and sorghum Sharma et al., 2005), and fruit crops like papaya and pineapple (Mahmud et al., 2018), are encouraged to grow and flourish by vermicompost. Applying vermicompost resulted in considerably higher growth and yield parameters for tomato plants, including stem diameter, plant height, marketable yield per plant, mean leaf number, and total plant biomass.

Plant growth and development are enhanced by the use of vermicompost, which improves the physical, chemical, and biological qualities of the soil. This, in turn, increases soil fertility, which in turn promotes plant growth and development. Vermicompost is a naturally occurring, slowly released source of plant nutrients that has been shown to increase plant dry weight and plant N uptake (Tomati et al., 1990).

However, a substantial amount of research indicates that while vermicompost has beneficial effects on plant growth and yield, these effects are not universal or consistent, and the strength of the effects reported in various studies varies greatly. Indeed, according to certain research, vermicompost may actually stunt plant growth or even kill them (Roberts et al., 2007; Lazcano et al., 2010c). The physical, chemical, and biological properties of vermicompost vary greatly depending on the original feedstock, the earthworm species used, the production process, and the age of vermicompost (Rodda et al., 2006, Warman and AngLopez, 2010). These factors may also affect the cultivation system into which vermicompost is incorporated.

Depending on the plant species or even variety under consideration, vermicompost's impacts might likewise vary greatly. This was seen in tomato plants, where three tomato varieties' germination, fruit shape, biomass allocation, and chemical characteristics were all affected differently when vermicompost was substituted for commercially fertilised potting media (Zaller, 2007).

Similar variation was noted in an experiment conducted by Lazcano et al. (2010a) to examine the effects of vermicompost and vermicompost extracts on the germination and early growth of six distinct maritime pine progenies. In this experiment, the maturity rate rose in three of the six pine progenies, fell in two, and remained unchanged in the other, when compared to the control group that did not receive vermicompost. It is reasonable to anticipate that distinct hybrids or genotypes of plants will react to vermicompost in different ways, given that genotype plays a significant role in determining the plant's ability to absorb nutrients, use them efficiently, and allocate resources.

In order to increase nutrient uptake, different genotypes may therefore promote root growth or alter root exudation patterns (Kabir et al. 1998; Cavani and Mimmo 2007). Each of these tactics will determine the establishment of distinct interactions with the microbial communities at the rhizosphere level. In fact, the rhizosphere microbial population of the various genotypes of sweet corn crops exhibited significant variations following the treatment of vermicompost (Aira et al. 2010).

This data makes it abundantly evident that vermicompost is a viable substitute for inorganic fertilisers in terms of stimulating plant growth. To keep consumers confident in this kind of fertiliser, more investigation is required into the precise processes and conditions that this organic substrate uses to promote plant development.

Vermicompost and soil health

Compacted soils are renewed and their ability to absorb water is enhanced when earthworms are present. The physical, chemical, and biological characteristics of organic matter and soil are altered by the feeding, burrowing, and casting behaviours of earthworms. Vermicompost typically has a higher nutrient profile than regular compost, as was previously mentioned. Vermicompost can actually improve the physical, chemical, and biological fertility of soil (Lim et al., 2015). Soils that have been enriched with vermicompost exhibit improved aeration, porosity, reduced bulk density, and increased water retention capacity.

Application of vermicompost greatly enhanced the chemical parameters of the soil, including pH, electrical conductivity, organic matter, and nutrient status; this resulted in improved plant growth and yield (Lim et al., 2015). Additionally, when casting, earthworms release a number of hormones, enzymes, and vitamins that encourage the activity of other helpful bacteria in the soil and enhance soil health.

Humus makes up a large portion of earthworm castings. Humus aids in the aggregation of soil particles, improving porosity and the soil's ability to hold water and facilitate aeration. Additionally, the humic acid in humus serves as a binding site for a number of plant nutrients, including calcium, iron, potassium, sulphur, and phosphorus. When the plants need these nutrients, they are released from their storage in the humic acid, where they are easily accessed (Adhikary 2012). Given that new castings have a higher moisture content and greater availability of nutrients, soil that has been consumed by earthworms may be even more conducive to plant growth (Hendriksen, 1997).

After applying vermicompost, Jeyabal and Kuppuswamy (2001) observed that rice stalk growth rose and soil fertility improved. According to Rose and Wood (1980), earthworm casts are typically credited with maintaining a healthy soil structure and enhancing the physical characteristics of the soil, such as water retention, infiltration, and erosion resistance. Vermicompost-amended soils can therefore retain more moisture and have better soil structure.

Due to the chemical and biological richness of earthworm casts, soils that have been intermingled with vermicompost have a higher capacity for cation exchange, a higher rate of humic acids and plant growth hormones, a higher population and activity of microorganisms, fewer root pathogens and soil-borne diseases, and an overall improvement in plant growth and yield (Arancon et al., 2003b, 2004a). Microorganisms found in worm casts may fix atmospheric nitrogen in amounts important to earthworm metabolism and as a source of nitrogen for plant growth, according to Lee (1992).

After adding vermicompost to the soil, Uee (1985) observed a six-fold increase in water infiltration and an 80% increase in hydraulic conductivity. According to Martin (1991), the percentage of macro-aggregates increased dramatically from 25.4 to 31.2% with the use of earthworm casts.

Maheswarappa's (1999) investigations showed that adding vermicompost to soil improved its organic carbon status, reduced bulk density, enhanced porosity and water-holding capacity, and boosted microbial populations and dehydrogenase activity. The average values of 48.2 and 11.9 g kg-1 soil, respectively, for the organic matter content of worm casts and surface soil have been reported to be about four times higher (Khang, 1994). Furthermore, according to Crossley (1988) and Bostom (1988), earthworms' contribution to N turnover in farmed soils ranged from 3 to 60 kg ha-1 year-1, improving the amount of N available to plants (Tiwari et al., 1989; Hullugalle and Ezumah, 1991).

Vermicompost and plant diseases

Vermicompost products have been shown to be useful as organic fertilisers and biological control agents, suppressing a number of plant diseases caused by foliar plant pathogens and soil-borne pests. Chemical pesticides were used excessively and repeatedly in conventional agriculture, which led to "biological resistance" in pests and crop diseases. Therefore, in order to prevent them from growing in high-yielding crops that are more susceptible to pests and diseases, far greater doses are now required (Patriquin et al., 1995). In order to compare the effectiveness of two distinct methods for inhibiting the growth of Sclerotium rolfsii-caused chickpea collar rot, two unconventional chemicals—$ZnSO_4$ and oxalic acid—as well as the bio-control agent Pseudomonas syringae—were used as seed coatings and foliar sprays, respectively. Vermicompost substitution was also used in this study. Vermicompost replacements considerably decreased chickpea mortality when compared to controls, but pre-inoculation with unconventional pesticides like foliar sprays against pathogens was far more effective in inhibiting growth (Sahni et al., 2008).

Applications of vermicompost prevented the development of tomato Fusarium wilt, which is caused by Fusarium lycopersici, and tomato late blight, which is caused by Phytophthora brassicae and Phytophthora nicotianae. Because earthworms stimulate soil microbial activity, they have a greater ability to inhibit plant diseases than aerobic compost. Numerous studies have been conducted on the suppressive impact of adding organic matter to soils, resulting in a noticeable decrease in plant parasitic nematode infestations. Scholarly papers regarding the suppressive effect of solid vermicomposts in comparison to OM and thermophilic compost additives on plant parasitic nematode numbers and outbreaks are scarce. Applications of solid vermicompost for nematode populations that parasite plants have been researched (Arancon et al., 2003).

Precautions during production

Precautions when producing vermicompost When creating vermicompost, find a shaded area and/or cover the unit with a shed or roof to protect the earthworms from the sun and rain. Water the gunny bags on a daily basis to keep the vermicompost bed or unit at the ideal moisture and temperature. Do not apply any type of fungicide or insecticide to the vermicompost bed. Vermicompost bedding material shouldn't be higher than three to five feet because doing so could raise the temperature inside the bed, cause an aeration issue, and cause the earthworms to die. Cover the vermicompost bed with a sieve to allow the earthworms to escape from the hens, birds, etc. (Kumar et al., 2018).

Limitations of vermicompost
- ➢ The process of vermicomposting takes time. Vermicompost is made by breaking down organic waste, which takes around six months.
- ➢ Higher maintenance is needed for vermicomposting compared to conventional composting.
- ➢ Because the vermicomposting pit must be kept at a cold enough temperature to support earthworm life, vermicompost may contain pests and diseases (Saha et al., 2022).

Conclusion

Vermicompost is not bad for the environment because it is an organic material. Additionally simple to use, even inexperienced small and marginal farmers can successfully prepare the vermicomposting procedure. Vermicompost has the potential to be a solution in the face of environmental deterioration and rising food demand. While its usage in agriculture alone would not be sufficient to meet demand for food, when combined with chemical fertiliser in an integrated manner, food production can be made sustainable. There is a low incidence of vermicompost adoption and a tendency for female farmers to be the only ones to do so. Vermicompost's potential hasn't yet been completely realised. Therefore, additional extension agents must be hired in order to inform farmers about vermicomposting and its advantages for attaining sustainability.

Acknowledgement

Authors are also thankful to The Principal, Shivanand College, Kagwad, Belagavi, Karnataka for providing digital facilities to review this work and also to the anonymous reviewer for thoroughly revising the manuscript.

Reference

Acevedo, I.C. and Pire, R. 2004. Effects of vermicompost as substrate amendment on the growth of papaya (*Carica papaya* L.). Interciencia, 29(5):274-279.

Adhikary, S. 2012. Vermicompost, the story of organic gold: A review. Agricultural Sciences, 3(7): 905-917.

Aira, M., Gómez-Brandón, M., Lazcano, C., Bååth, E. and Domínguez, J. 2010. Plant genotype strongly modifies the structure and growth of maize rhizosphere microbial communities. Soil Biology and Biochemistry 42(12):2276-2281.

Allen, J. 2016. Vermicompost Guide H-164. NM State University, Mexico. pp-1-4.

Arancon, N.Q., C.A. Edwards, P. Bierman, C. Welch, and Metzger, J.D. 2004a. The influence of vermicompost applications to strawberries: Part I. Effects on growth and yield. Biores. Technol., 93:145-153.

Arancon, N.Q., Galvis, P., Edwards, C. and Yardim, E. 2003. The trophic diversity of nematode communities in soils treated with vermicompost: The 7th International Symposium on Earthworm Ecology·Cardiff·Wales 2002. Pedobiologia, 47(5-6):736-740.

Arancon, N.Q., S. Lee, C.A. Edwards, and Atiyeh, R.M. 2003b. Effects of humic acids and aqueous extracts derived from cattle, food and paper-waste vermicomposts on growth of greenhouse plants. Pedobiologia, 47: 744-781.

Bansal, S., and Kapoor, K.K. 2000. Vermicomposting of crop residues and cattle dung with *Eisenia foetida*. Bioresour. Technol 73(2):95 98.

Bostom. 1988. Pedobiologia, 3:379-388.

Cavani, N. and Mimmo, T. 2007. Rhizodeposition of *Zea mays* L. as affected by heterosis. Archives of Agronomy and Soil Science, 53:593-604.

Chowdeppa, P., Biddappa, C.C. and Sujatha, S. 1999. Effective recycling of organic waste in arecanut (*Areca ctechu* L.) and cocoa (*Theombrome cacoa* L.) plantation through vermicomposting. Indian J Agric Sci., 69:563-566.

Crossley. 1988. Applied Eco, 26:505-520.

Dominguez, J., and Edwards, C.A. 2004. Vermicomposting organic wastes: A review. *In*: Soil Zoology for Sustainable Development in the 21st Century (Shakir, S.H., Mikhail, W.Z.A., Eds).

Edwards, C.A. 1998. The use of earthworm in the breakdown and management of organic waste. *In:* Earthworm Ecology. ACA Press LLC, Boca Raton, FL, pp. 327-354.

Edwards, C.A. and Burrows, I. 1998. The potential of earthworm composts as plant growth media. In: Edwards CA, Neuhauser E, editors. Earthworms in Waste and Environmental Management.: SPB Academic Press, The Hague, the Netherlands. pp. 21-32.

Hendriksen. 1997. Biotech. Bioeng. 23: 1812-1997.

Hullugalle, N.R. and Ezumah, H.C. 1991. Agric., Ecosystems and Environ, 35:55-63.

Jeyabal, G., and Kuppuswamy. 2001. Recycling of organic wastes for the production of vermicompost and its response in rice-legume cropping system and soil fertility. Eur. J. Agron, 15 (3):153 170.

Kabir, Z., O'Halloran, I. P., Fyles, J. W. and Hamel, C. 1998. Dynamics of the mycorrhizal symbiosis of corn (*Zea mays* L.): effects of host physiology, tillage practice and fertilization on spatial distribution of extra-radical mycorrhizal hyphae in the field. Agriculture, Ecosystems and Environment, 68:151-163.

Kalantari, S., Hatami, S., Ardalan, M.M., Alikhani, H.A. and Shorafa, M. 2010. The effect of compost and vermicompost of yard leaf manure on growth of corn. African Journal of Agricultural Research, 5:1317-1323.

Khang, B.T. 1994. Soil Fert. Soils, 18: 193-199.

Kiehl, J.C. 2001. Producao de composto organico e vermicomposto. Informe Agropecua rio, Belo Horizonte, 22 (212):40-52.

Kumar, A., Bhanu Prakash, C.H., Brar, N.S. and Kumar, B. 2018. Potential of Vermicompost for Sustainable Crop Production and Soil Health Improvement in Different Cropping Systems. Int.J.Curr.Microbiol.App.Sci, 7(10):1042-1055.

Lazcano, C. and Domínguez, J. 2010c. Effects of vermicompost as a potting amendment of two comercially-grown ornamental plant species. Spanish Journal of Agricultural Research, 8(4):1260-1270.

Lazcano, C., Sampedro, L., Zas, R. and Domínguez, J. 2010a. Vermicompost enhances germination of the maritime pine (*Pinus pinaster* Ait.). New Forest, 39:387-400.

Lee, K.E. 1992. Soil Bio Biochem. 24: 1765-1771.

Lim, S.L., T.Y. Wu, P.N. Lim, and Shak, K.P. 2015. The use of vermicompost in organic farming: overview, effects on soil and economics. Journal of Science and Food Agriculture, 95(6):1143-1156.

Lores, M., M. Gómez-Brandón, D. Pérez-Díaz and Domínguez, J. 2006. Using FAME profiles for the characterization of animal wastes and vermicomposts. Soil Biology and Biochemistry, 38, 2993-2996.

Mahewarappa, H.P., Nanjappa, H.V. and Hegde, M.R. 1999. Influence of organic manures on yield of arrowroot, soil physico-chemical and biological properties when grown as intercrop in coconut garden. Annals of Agricultural Research, 20:318-323.

Mahmud, M., Abdullah, R. and Yaacob, J.S. 2018. Effect of vermicompost amendment on nutritional status of sandy loam soil, growth performance, and yield of pineapple (*Ananas comosus* var. MD2) under field conditions. Agronomy, 8(9):183-189.

Martin, A. 1991. Biology and Fertility of Soils. 11: 234-238.

Nada, W.M., Van Rensburg, L. and Claassens, S. 2011. Communications in soil science and plant analysis effect of vermicompost on soil and plant properties of coal spoil in the Lusatian region (Eastern Germany). Communications in Soil Science and Plant Analysis, 42(16):1945-1957.

Patriquin, D.G., Baines, D. and Abboud, A. 1995. Diseases, pests and soil fertility. In: Soil Management in Sustainable Agriculture. Wye College Pres, UK. pp. 161-174.

Reddell, P. and Spain, A.V. 1991. Earthworms as vectors of viable propagules of mycorrhizal fungi. Soil Biology and Biochemistry. 23(8):767-774.

Roberts, P., Jones, D.L. and Edwards-Jones, G. 2007. Yield and vitamin C content of tomatoes grown in vermicomposted wastes. Journal of the Science of Food and Agriculture, 87:1957-1963.

Rodda, M.R.C., Canellas, L.P., Façanha, A.R, Zandonadi, D.B., Guerra, J.G.M., De Almeida, D.L. and De Santos, G.A. 2006. Improving lettuce seedling root growth and ATP hydrolysis with humates from Vermicompost. II- Effect of Vermicompost source. Revista Brasileira de Ciencia do Solo, 30:657-664.

Rose, C.J., and Wood, A.W. 1980. Some environmental factors affecting earthworm populations and sweet potato production in Tari basin, Papua New Guniea Highlands. Papua New Guinea Agric. J., 31: 1-13.

Rostami, R., A. Nabaei, and Eslami, A. 2009. Survey of optimal temperature and moisture for worms' growth and operating vermicompost production of food wastes. Health and environment, 1 (2): 105-112.

Saha, P., Barman, A. and Bera, A. 2022. Vermicomposting: A step towards sustainability. Sustainable Crop Production.

Sahariah, B., Goswami, L., Kim, K., Bhattacharyya, P. and Sundar, S. 2015. Metal remediation and biodegradation potential of earthworm species on municipal solid waste: A parallel analysis between Metaphireposthuma and *Eisenia fetida*. Bioresource Technology, 180:230-236.

Sahni, S., Sarma, B.K. and Singh, K.P. 2008. Management of *Sclerotium rolfsii* with integration of non-conventional chemicals, vermicompost and *Pseudomonas syringae*. World Journal of Microbiology and Biotechnology, 24(4):517-522.

Sharma, S., Pradhan, K., Satya, S. and Vasudevan, P. 2005. Potentiality of earthworms for waste management and in other uses—A review. Journal of American Science, 1(1):4-16.

Subbulakshmi, G. and Thiruneelakandan R. 2011. Vermicomposting is valiant in vandalizing the waste material. International Journal of Plant, Animal and Environmental Sciences. 1(3):134-141.

Tharmaraj, K., Ganesh, P., Kolanjinathan, K., Suresh, K.R. and Anandan, A. 2011. Influence of vermicompost and vermiwash on physico chemical properties of rice cultivated soil. Current Botany, 2(3):18-21.

Tiwari, K.N. 1989. Fertilizer management in cropping system for increased efficiency. Fertilizer News, 25(3):3-20.

Tomati, U., and Galli, E. 1995. Earthworms, soil fertility and plant productivity. Acta Zoologica Fennica, 196: 11-14.

Tomati, U., Galli, E., Grappelli, A. and Di Lena, G. 1990. Effect of earthworm casts on protein synthesis in radish (*Raphanus sativum*) and lettuce (*Lactuga sativa*) seedlings. Biology and Fertility of Soils, 9(4):288-289.

Ushakumari, K., Prabhakumari, P. and Padmaja, P. 1999. Efficiency of vermicomposts on growth and yield of summer crop okra (*Abelmoschus esculentus* Moench). J. Trop. Agric., 37: 87-88.

Vasanthi, D., and Kumarasamy, K. 1999. Efficacy of vermicompost to improve soil fertility and rice yield. Journal Indian Society of Soil Sciences, 42(2):268-272.

Warman, P.R. and AngLopez, M.J. 2010. Vermicompost derived from different feedstock's as a plant growth medium. Bioresource Technology, 101:4479-4483.

Zaller, J.G. 2007. Vermicompost as a substitute for peat in potting media: Effects on germination, biomass allocation, yields and fruit quality of three tomato varieties. Scientia Horticulturae, 112:191-199.

YOUR KNOWLEDGE HAS VALUE

- We will publish your bachelor's and
  master's thesis, essays and papers

- Your own eBook and book -
  sold worldwide in all relevant shops

- Earn money with each sale

Upload your text at www.GRIN.com
and publish for free